THE SUN

Rebecca Woodbury, Ph.D., M.Ed.

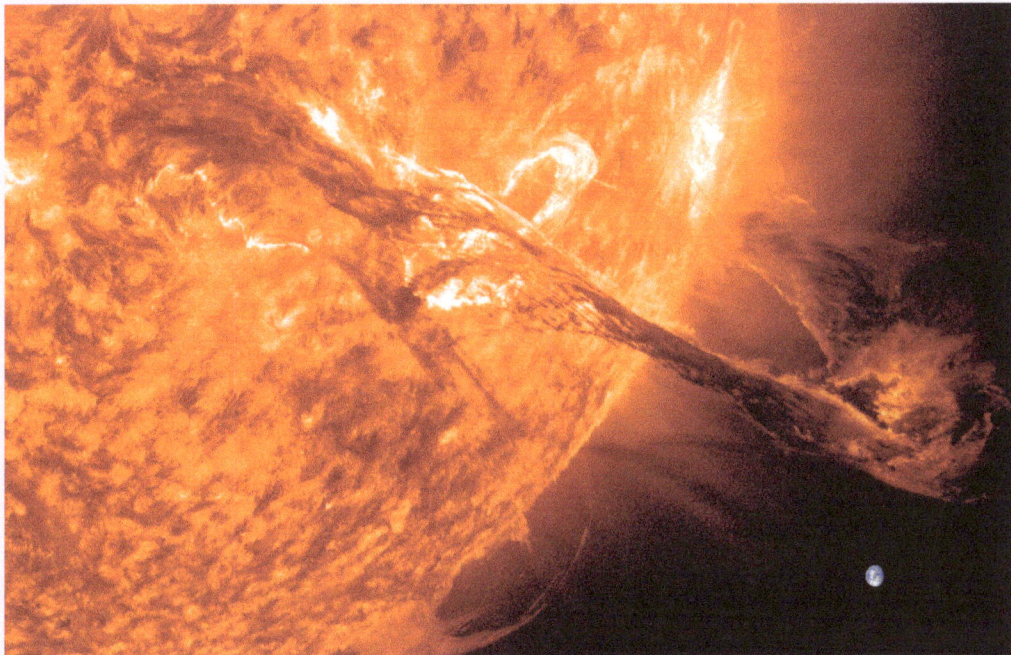

Gravitas Publications Inc.

THE SUN

Illustrations: Janet Moneymaker

The Sun
ISBN 978-1-950415-41-0

Published by Gravitas Publications Inc.
Imprint: Real Science-4-Kids
www.gravitaspublications.com
www.realscience4kids.com

RS4K

Photo credits: Cover & Title Pg: NASA/GSFC/SDO; Above & P.13. NASA; .5. NASA; P.19. By yod67, AdobeStock

You can see the Sun rise in the morning and move across the sky during the day.

When you go outside in the summertime, you can feel the warmth from the Sun.

Summertime is fun!

The Sun is much larger than Earth. It is so huge that a million Earths would fit inside.

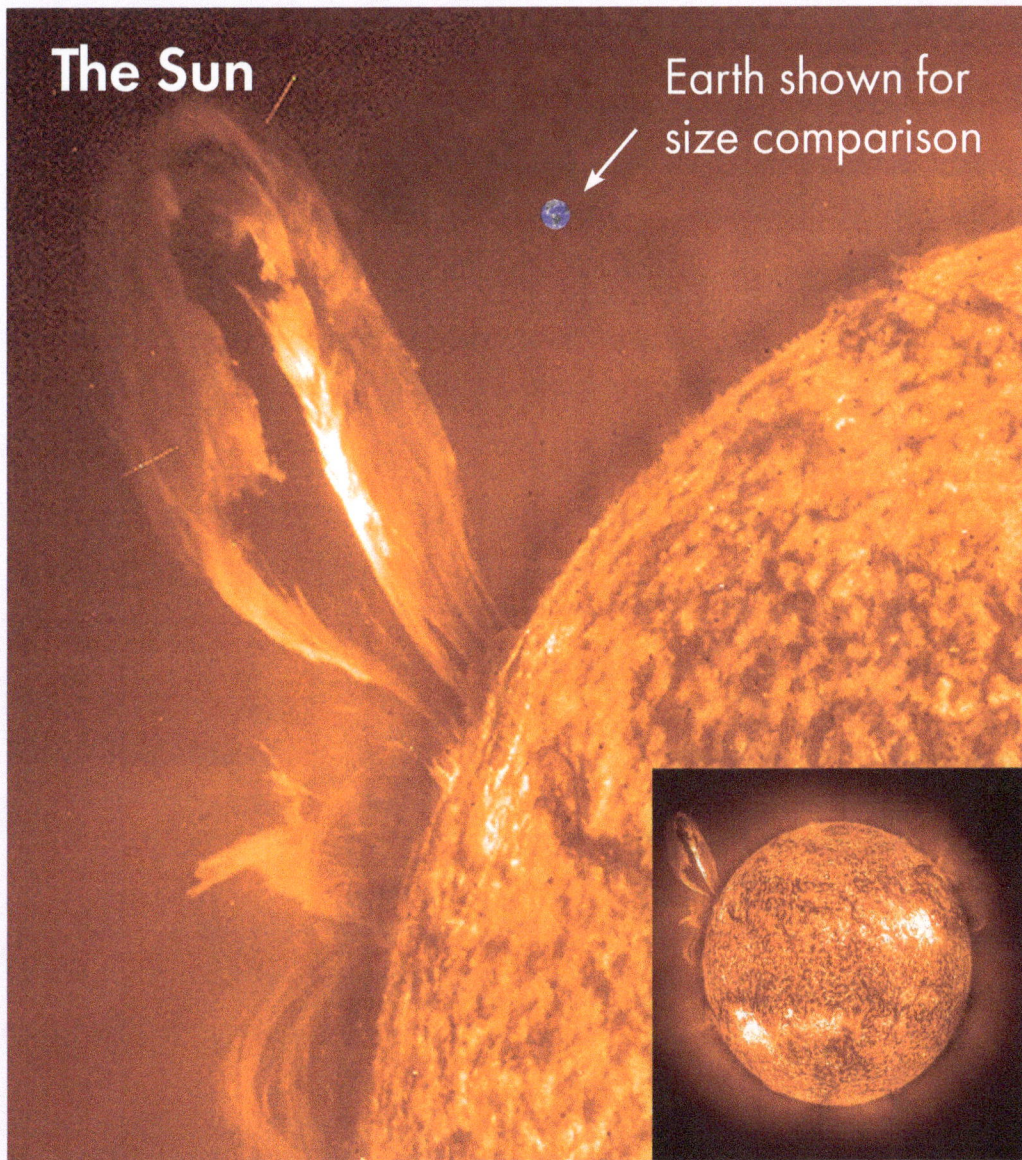

The Sun

Earth shown for size comparison

The Sun is not a **planet** or a **moon**. The Sun is a **star!**

A star is an object in space where millions of atoms are wiggling and bumping into each other so many times that BIG changes are occurring, which is why we feel warm and get light from the Sun.

Because there is so much change, the measurement we get — energy — is a very big number.

Wow! Stars are suns!

Review: ENERGY

Energy is a measure of the amount of change that happens.

When something changes a small amount, we measure a small energy change.

When something changes a big amount, we measure a BIG energy change.

The Sun is not made of rocks and soil like Earth. Instead, the Sun is made of **helium** gas and **hydrogen** gas.

I am Hydrogen.

Hi! I am Mouse.

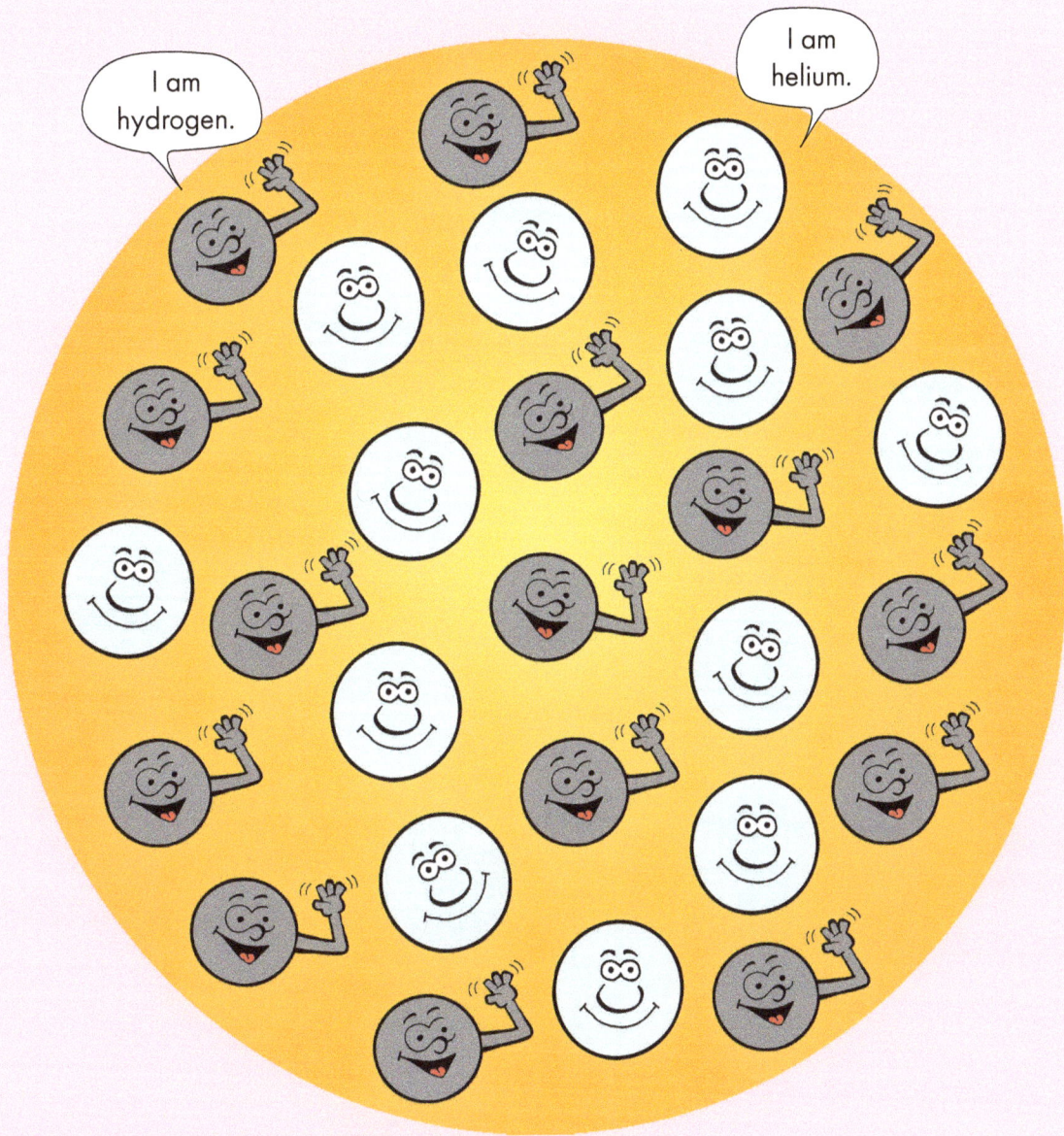

Inside the Sun, hydrogen atoms and helium atoms rearrange in a nuclear reaction. This rearranging causes a HUGE amount of change as the Sun's atoms wiggle much faster — so fast that the Sun shines and makes us feel warm.

Did you know that atoms are little building blocks that make up everything we touch, taste, smell, and see?

Yes! And atoms are too tiny to see with our eyes.

The Sun

NUCLEAR REACTION!

Hydrogen atoms combine

A helium atom is made

Gives off light!

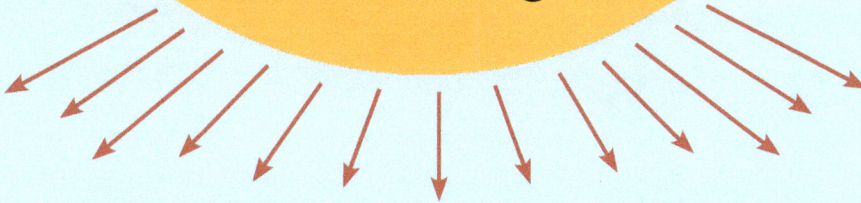

Allows us to see the world and
MAKES US FEEL WARMER.

Because the atoms inside the Sun wiggle so fast, the Sun's temperature is a **very BIG number!** The Sun's temperature is so high that you can't get near it even with the best spaceship!

Sunlight makes it possible for
living things to exist on Earth.

Hooray for sunlight!

Earth is just the right temperature
for living things to exist because
Earth is just the right distance
from the Sun.

Sunlight gives plants the ability to grow and to make sugar for their food. The sugar made by plants also feeds all the animals.

Many astronomers are now studying the Sun. They are learning more about how the Sun works and how it affects Earth.

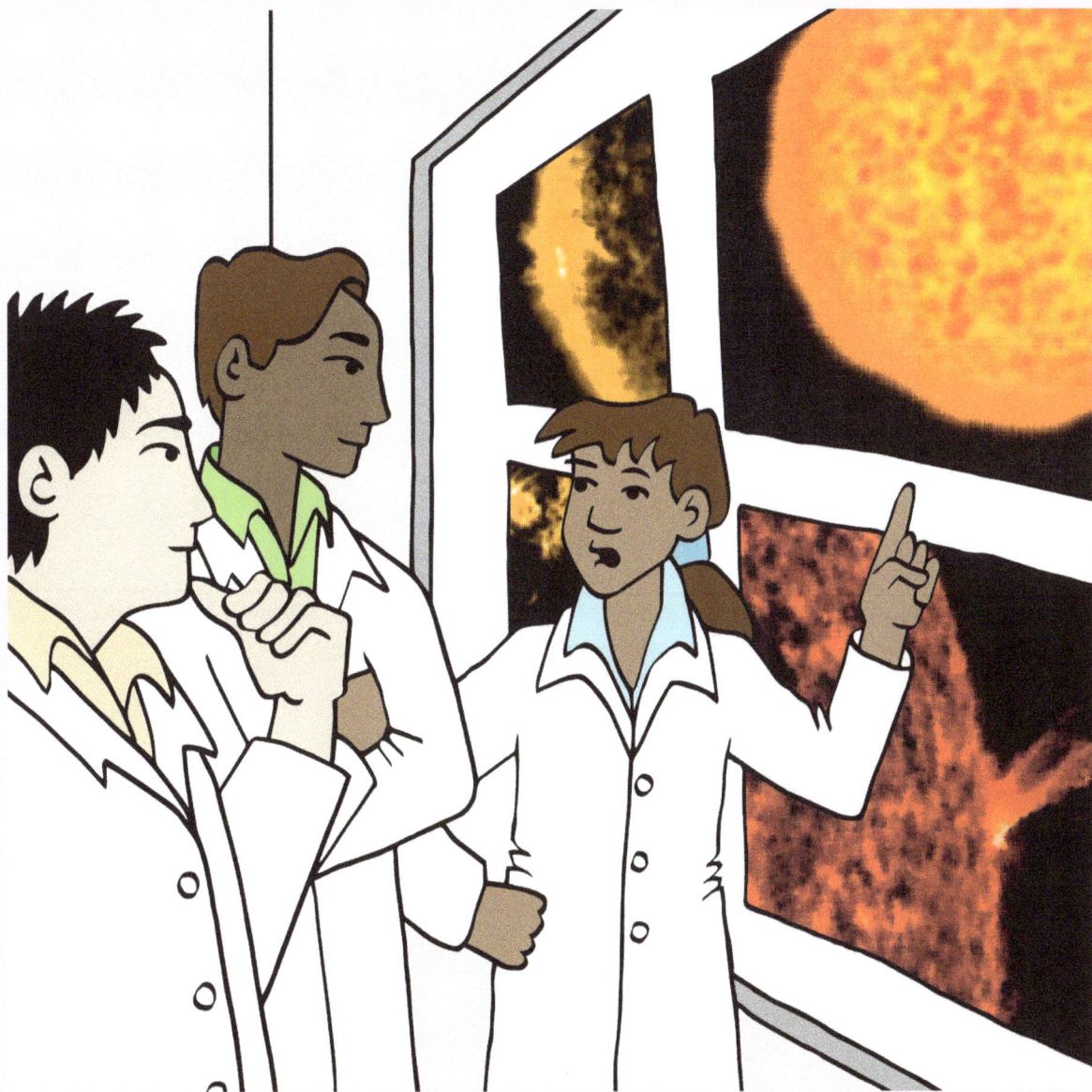

How to say science words

astronomer   (uh-STRAH-nuh-muhr)

atom   (AA-tuhm)

Earth   (ERTH)

energy   (EN-uhr-jee)

heat   (HEET)

helium   (HEE-lee-uhm)

hydrogen   (HIY-druh-juhn)

light   (LIYT)

nuclear reaction   (NOO-klee-uhr
                          ree-AAK-shuhn)

planet   (PLAA-nuht)

science   (SIY-uhnss)

space   (SPAYSS)

star   (STAHR)

www.ingramcontent.com/pod-product-compliance
Lightning Source LLC
Chambersburg PA
CBHW040153200326
41520CB00028B/7587

References

Chapter 1:

"Vagaries of a Financier." New York Times, February 18th, 1894

"Mission Preparation and Pre-Launch Operations." Retrieved at: http://science.ksc.nasa.gov/shuttle/technology/sts-newsref/stsover-prep.html. June 11, 2013.

"This is a timeline for a Shuttle Astronaut." Retrieved at: http://space flight.nasa.gov/living/timeline_shuttle_24fps.swf. June 11, 2013.

Chapter 2:

Cappozzoli, Thomas K. (1995, Dec). "Resolving conflict within teams." Journal for Quality and Participation. v18n7, p. 28-30.

Lencioni, Patrick. The Five Dysfunctions of a Team. Jossey-Bass Press, 2002.

Tuckman, Bruce W. (1965) "Developmental sequence in small groups." Psychological Bulletin, 63, 384-399.

"Unhealthy conflict hurts workers." Retrieved at: http://www.seattlepi.com/business/article/Workplace-Coach-Unhealthy-conflict-hurts-workers-1254597.php. June 11, 2013

Chapter 3:
"NASA trying to avoid layoffs, congressman says." Compiled from Times wires, © St. Petersburg Times; published March 17, 2003

"Space Shuttle Columbia disaster." Retrieved from: Wikipedia, the free encyclopedia. June 11, 2013.

Lipartito, Kenneth and Orville R. Butler. A history of the Kennedy Space Center. University Press of Florida, 2007.

Chapter 4:
Welch, Jack. Winning. Harper Collins Publishing, 2005.

Chapter 5:
Cabbage, Michael. "After $273 Million, NASA Scraps Launch-Control Computer System Project." McClatchy - Tribune Business News. September 17, 2002.

References

Lipartito, Kenneth and Orville R. Butler. A history of the Kennedy Space Center. University Press of Florida, 2007.

"Sun machines power shuttle launch system." ServerWorld 13.October 1999 p.6-7.

Chapter 6:
"How to Improve Your Hiring Practices." Inc. Magazine. April 1, 2010.

Fernández-Aráoz, Groysberg, and Nohria.The Definitive Guide to Recruiting in Good Times and in Bad. Harvard Business Review, May 2009.

Chapter 7:
Johnston, Don. "Fear not the Expert." Proceedings of the 2012 International Lean and Six Sigma Conference.

Useem, Michael. "The Go Point." Crown Publishing Group, 2006.

Chapter 8:
3 Spacewalking Astronauts Capture Satellite. The Washington Post. May 14, 1992

Sawyer, Kathy. "Relieved astronauts describe improvising satellite's rescue." The New York Times. May 16, 1992.

Stewart, Robert. "3 Astronauts Grab Marooned Satellite in a Dramatic Rescue." Los Angeles Times. May 14, 1992

Chapter 9:
Feldstein, Dan & Bill Murphy. "Shuttle chief admits to staff attrition." The Houston Chronicle. March 7, 2003.

Pae, Peter. "NASA Lifts Palmdale Off Shuttle Maintenance List." Los Angeles Times. February 2, 2002.

St. John, Paige. "Calif., Fla. lawmakers spar on shuttle maintenance." Gannett News Service. February 28, 2002.

Chapter 10:
Fowler, Wallace T. "An Introduction to Space Mission Planning." Retrieved at: http://design.ae.utexas.edu/mission_planning/

References

Lipartito, Kenneth and Orville R. Butler. A history of the Kennedy Space Center. University Press of Florida, 2007.

Author's Note
Deming, W. Edwards. Out of the Crisis. First MIT Press. Cambridge, MA, 2000.

Sheep & Shuttles

About The Authors

Keith Ruehl:

Keith and his family call the mountains of western North Carolina home. Over the past 15 years the family's hobby farm experiences have provided many valuable and entertaining lessons. Keith earned his undergrad at Northwood University in Business Administration and later earned an MBA from Nova Southeastern University. Keith started his own business at age 23, and after 10 years of successful growth sold it to a national firm where he was subsequently employed for 15 years in various senior management leadership positions. He currently is a Managing Partner for Sustainable Resort Institute, an adjunct professor at Mars Hill University, and a Managing Partner for the business consulting firm Your Answer Key. He is a volunteer firefighter and enjoys martial arts and scuba diving.

Anthony Delmonte:

Tony and his wife, Trudy, live in Central Florida. He spent eighteen years in senior management working on the Shuttle program at Kennedy Space Center, before retiring to be an adjunct professor for an aeronautical university and serve as Managing Partner for the business consulting firm Your Answer Key. Tony earned his Bachelors in Management at Indiana University, a Masters in Aeronautical Science at Embry-Riddle Aeronautical University, and a Doctorate in Business Administration from Nova Southeastern University. In his spare time Tony enjoys traveling, golfing and solving crossword puzzles.

www.ingramcontent.com/pod-product-compliance
Lightning Source LLC
Chambersburg PA
CBHW032007190326
41520CB00007B/384